ALGORITHMIC
WARFARE
THE RISE OF AUTONOMOUS
WEAPONS
JOSH LUBERISE
FORTIS NOVUM MUNDUM

Algorithmic Warfare: The Rise of Autonomous Weapons

Josh Luberisse

Published by Fortis Novum Mundum, 2023.

ALGORITHMIC WARFARE: THE RISE OF AUTONOMOUS WEAPONS

First edition. February 20, 2023.

ISBN: 979-8223471912

Written by Josh Luberisse.

Table of Contents

Algorithmic Warfare: The Rise of Autonomous Weapons

By

Josh Luberisse

Table of Contents

Preface

In recent years, the development and deployment of autonomous weapons have become an increasingly controversial and polarizing topic. On one hand, proponents of autonomous weapons argue that these systems have the potential to increase military efficiency, reduce the risk to human lives, and enhance national security. On the other hand, opponents of autonomous weapons express concern about the ethical, moral, and legal implications of allowing machines to make decisions that could result in loss of human life and the potential risks of unintended consequences.

This book aims to provide a comprehensive overview of the current state of autonomous weapons development, deployment, and use in modern warfare. It covers the historical, ethical, legal, technological, and political aspects of algorithmic warfare, and explores the potential risks and benefits of autonomous weapons. The book also provides a critical examination of the challenges and opportunities associated with the operationalization of artificial intelligence in warfare and the impact of autonomous weapons on society.

This book is intended to serve as a reference and resource for policymakers, military leaders, researchers, and anyone interested in understanding the complex and rapidly evolving field of algorithmic warfare. We hope that the insights and analysis presented in this book will help foster informed debate and contribute to the development of responsible and ethical policies for the use of autonomous weapons in the future.

Prologue

The sky is lit up with flashes of light and the ground shook with a deafening roar as the bombs fell from the sky. This is a scene that has been repeated countless times throughout history, but this time it is different. This time, the bombs will not be dropped by human pilots, but rather by autonomous drones. As someone who grew up in the poorest nation in the Western Hemisphere, I have seen the worst of warfare firsthand.

Writing this book has opened my eyes to the advancements in technology that has led to the development of unmanned aerial vehicles and the widespread use of artificial intelligence in military operations. And as a history buff, I have read countless manuals on the history of warfare, including the role of technology in changing the way wars are fought. But this is something new. The use of autonomous weapons is not just the next step in the evolution of warfare; it is a fundamental shift in the way wars are fought. The technology has advanced to the point where autonomous machine are making life and death decisions, and this will fundamentally change the nature of war.

It is clear to me now that this new era of warfare will have far-reaching consequences, not just for the military, but for society as a whole. As the capabilities of autonomous weapons continues to grow, the line between man and machine is becoming increasingly blurred, and the ethical and legal implications of their use are becoming more complex.

It is this realization that led me to write this book. I wanted to explore the history, the current state, and the future of autonomous weapons. I wanted to examine the potential benefits and risks of their use, and the ethical and legal considerations that policymakers and military leaders must take into account when deciding how to employ them.

This book is not just for military professionals or academics; it is for anyone who wants to understand the impact of autonomous weapons on our society and our future. It is a call to action, urging us to engage in a critical dialogue about the role of technology in warfare and to consider the implications of our decisions.

The age of autonomous weapons is upon us, and it is up to us to determine how we will use this technology. This book is a starting point for that discussion.

Unit I. Introduction

The world of warfare is undergoing a profound transformation with the advent of algorithmic warfare and autonomous weapons. The rapid advancement of technology has enabled the development of systems that can operate with increasing levels of autonomy, creating new and complex ethical, legal, and strategic dilemmas. This book explores the rise of algorithmic warfare, its current state, and its implications for the future of conflict.

The term "algorithmic warfare" refers to the use of algorithms and artificial intelligence in military operations and decision-making. Autonomous weapons, also known as killer robots, are a specific subset of algorithmic warfare, designed to operate independently with minimal human intervention.

The development of autonomous weapons raises important questions about the future of warfare and the role of humans in making decisions about the use of military force. While these weapons offer significant operational benefits, such as faster reaction times and reduced risk to human life, they also pose new and complex ethical, legal, and strategic challenges.

The aim of this book is to provide a comprehensive overview of algorithmic warfare and autonomous weapons, exploring their development, current state, and future implications. Through a combination of historical analysis, expert insights, and case studies, this book will provide readers with a clear understanding of the key issues surrounding algorithmic warfare and autonomous weapons.

In the following chapters, we will examine the historical development of algorithmic warfare and the evolution of military technology. We will explore the current state of autonomous weapons, their use in contemporary conflicts, and the trends and predictions for the future. The ethical and legal implications of algorithmic warfare will be discussed, along with the risks and benefits of autonomous weapons and the international responses to this new form of warfare. Finally, we will examine the future of algorithmic warfare and its impact on society and offer insights into how best to prepare for a world where machines play an increasingly central role in military operations.

Objectives

THE PURPOSE OF THIS book is to provide a comprehensive overview of algorithmic warfare and its implications for the future of conflict. As the use of algorithms and artificial intelligence in military operations continues to grow, it is essential to understand the implications of these technologies for the conduct of warfare and the role of humans in the use of military force.

This book explores the historical development of algorithmic warfare, the current state of the field, and its future implications. Through a combination of historical analysis, expert insights, and case studies, this book provides a comprehensive examination of the key issues surrounding algorithmic warfare and autonomous weapons.

This book aims to answer questions such as:

- What is algorithmic warfare and how has it evolved over time?
- What are the key benefits and risks of algorithmic warfare and autonomous weapons?
- What are the ethical and legal implications of algorithmic warfare and autonomous weapons?

- How are nations and international organizations responding to the rise of algorithmic warfare and autonomous weapons?
- What is the future of algorithmic warfare and what are the implications for the future of conflict?

This book is intended for a wide audience, including military planners, policymakers, security experts, academics, and anyone interested in the future of warfare and military strategy. Through a clear and accessible examination of the key issues, this book will provide readers with a comprehensive understanding of algorithmic warfare and its implications for the future of conflict.

Defining Algorithmic Warfare

ALGORITHMIC WARFARE, also known as algorithmic conflict, refers to the use of algorithms and artificial intelligence in military operations and decision-making. The integration of these technologies into military systems has enabled a new era of warfare, characterized by increased automation, faster decision-making, and reduced human involvement in the use of military force.

The development of algorithmic warfare has been driven by the rapid advancement of technology and the increasing use of digital systems in military operations. Algorithms and artificial intelligence have been used to automate routine tasks, improve situational awareness, and enhance decision-making processes. These technologies are now being integrated into weapons systems, allowing for increasing levels of autonomy in the use of military force.

The current usage of the term 'Algorithmic Warfare' originates from Project Maven launched in April 2017 by the U.S. Department of Defense. This project aimed to train AI systems using advanced computing techniques so that they can autonomously analyze massive amounts of data collected by US drones.

According to the Pentagon, the goal of the project was to create and implement "computer-vision algorithms needed to help military and civilian analysts encumbered by the sheer volume of full-motion video data that DoD collects every day in support of counterinsurgency and counterterrorism operations." The Pentagon had hoped to implement its first batch of algorithms into "warfighting systems" by the end of 2017.

One of the key benefits of algorithmic warfare is improved operational efficiency as autonomous weapon systems can operate with faster reaction times, greater accuracy, and reduced risk to human life. This has the potential to revolutionize the way wars are fought, offering new operational capabilities and opportunities for military planner.

However, the use of algorithmic warfare also raises significant ethical, legal, and strategic dilemmas. Autonomous weapons, for example, raise questions about the accountability of machines for the use of military force, and the potential for autonomous systems to act in ways that are contrary to international law or ethical norms.

The rise of algorithmic warfare has prompted international discussions about the regulation of autonomous weapons, with many countries and organizations calling for a ban on killer robots. At the same time, there is growing recognition of the importance of investing in these technologies to maintain a competitive military edge in an increasingly complex and rapidly changing security environment.

Algorithmic warfare represents a new and rapidly evolving domain of military operations, with significant implications for the future of warfare. Understanding the development, use, and impact of these technologies is essential for military planners, policymakers, and anyone interested in the future of conflict.

Defining Autonomous Weapon Systems (AWS)

THE DEVELOPMENT OF autonomous weapon systems (AWS) has been a hot topic in the field of military technology, with various countries exploring their use in modern warfare. However, before discussing the implications and potential risks of these systems, it is important to first define autonomous weapon systems is and how they work.

An autonomous weapon system (AWS) is a weapon system that can select and engage targets without human intervention. This means that these systems have varying degrees of autonomy and can operate in different ways depending on the level of human involvement in their operation.

There are several ways to classify the level of autonomy in autonomous weapon systems. One common classification is based on the level of human involvement in the system's decision-making process. The levels of human involvement are "in the loop," "on the loop," and "out of the loop."

"Human in the loop" autonomous weapon systems refers to weapon systems that require a human operator to make the final decision before taking any action. These systems provide suggestions or recommendations to the operator, who then decides whether to engage the target or not. According to the Department of Defense Directive (DODD) 3000.09, an “human in the loop” is a semi-autonomous weapon system that will “only engage individual targets or specific target groups that have been selected by a human operator.” Examples of “in the loop” weapon system include “fire and forget” weapons such as guided missiles.

“Human on the loop” autonomous weapon systems refer to weapon systems in which human operators can monitor and halt the weapon's target engagement.

"Human out of the loop" autonomous weapon systems refer to fully autonomous weapon systems that can operate without any human intervention. These systems are designed to identify and engage targets on their own, without requiring any input from a human operator.

Another way to classify AWS is based on the degree of supervision provided to the system. The three main classifications are supervised, unsupervised, and semi-supervised autonomous systems.

Supervised autonomous systems require human supervision throughout their operation. The human operator is responsible for programming the system's decision-making algorithm, and the system operates within the parameters set by the operator.

Unsupervised autonomous systems, on the other hand, do not require human supervision. These systems are programmed with a set of parameters and can operate without human intervention, making decisions and engaging targets based on the information gathered by the system's sensors.

Semi-supervised autonomous systems are a combination of supervised and unsupervised systems. These systems are designed to operate without human intervention, but the human operator has the ability to intervene if necessary.

Regardless of the level of autonomy and supervision, all autonomous weapon systems consist of three main components: sensors, a decision-making system, and munitions. The sensors gather information about the environment and potential targets, and the decision-making system processes that information to determine if and when to engage a target. The munitions are the actual weapons that the system uses to engage targets, whether it is a missile, a drone, or another type of weapon.

As the development and deployment of autonomous weapon systems continues to grow, it is crucial to understand the different levels of autonomy and the various components that make up these systems. By gaining a better understanding of how these systems work, policymakers and military leaders can make informed decisions about their use and potential risks.

Autonomous Weapons Systems Engagement Loop

The Autonomous Weapons Systems Engagement Loop (AWSEL) is a term used to describe the interaction between an autonomous weapon system (AWS) and its targets. Understanding how this loop works is critical to understanding how AWS operates and the potential benefits and risks associated with their use. Autonomous Weapons Systems can be classified in terms of the level of human involvement in their operations. These range from fully autonomous systems, where humans are not involved at all, to semi-autonomous systems, where humans play a significant role in decision-making. Based on this level of human involvement and oversight, AWS can be further categorized as either supervised, unsupervised, or semi-supervised.

As mentioned in the previous unit autonomous weapon system can be designed to operate with different levels of human involvement. Humans "in the loop" (HITL) weapon systems refer to an autonomous weapon system that receives guidance from and is controlled remotely by a human operator, who is responsible for monitoring and making decisions about the system's actions. In this scenario, the human is effectively operating the autonomous weapon system, and the autonomous weapon system cannot operate without the human's input. Example of this include the infamous Predator and its successor the MQ-9 Reaper drones used by the US military in Iraq, Afghanistan.

On the loop (OTL) refers to an autonomous weapon system that receives input from a human operator, but the human is not actively making decisions about the system's actions. Instead, the operator is providing guidance to the autonomous weapon system, which is then making decisions based on the input it receives. In this scenario, the human is not controlling the autonomous weapon system, but is rather acting as a supervisor, ensuring that the system is performing its tasks in a safe and effective manner.

Out of the loop (OOTL) refers to a level of human involvement in which the autonomous weapon system is operating independently, without any input from a human operator. In this scenario, the human has no direct control over the autonomous weapon system, and the system is making decisions on its own.

Lethal Autonomous Weapons (LAWS) are perhaps the most well known 'human out of the loop' autonomous weapon systems. The Department of Defense Directive (DODD) 3000.09, which establishes U.S. policy on autonomy in weapons systems, defines LAWS as "weapon system[s] that, once activated, can select, and engage targets without further intervention by a human operator."

The level of human involvement in autonomous weapon systems can also be categorized by the level of supervision. Supervised autonomous systems (SAS) are those in which humans are closely monitoring the system's behavior and providing guidance as needed. Unsupervised autonomous systems (UAS) are those that operate with no human oversight, and make decisions based solely on the data they receive. Semi-supervised autonomous systems (SSAS) are those that operate with a combination of human oversight and autonomous decision-making. In these systems, humans monitor the system's behavior, but the system is also able to make decisions on its own, within a set of predetermined rules.

The decision to use an autonomous weapon system with or without human involvement can have significant implications for its effectiveness and safety. A system with human control may be slower to react, but may also be more precise in its actions, as the human operator can take into account a wider range of factors. An unsupervised system, on the other hand, may be faster and more efficient in its actions, but may also be more prone to errors or unintended consequences.

In general, it is important to carefully consider the level of human involvement and supervision when designing and implementing autonomous weapon systems. While fully autonomous systems may offer some benefits, they also come with significant risks, and it is important to ensure that appropriate safety measures are in place to prevent unintended consequences. By understanding the different categories of autonomous weapon systems and their engagement loops, policymakers and military leaders can make informed decisions about how best to utilize these technologies while minimizing risks and maximizing benefits.

Homing munitions and Loitering munitions

Homing and loitering munitions are two types of weapons that have become increasingly prevalent in modern warfare. While both of these types of weapons can be autonomous, they differ in their primary function and design.

Homing munitions are typically designed to engage a specific target, whether that is an aircraft, a ground vehicle, or a building. These weapons are often guided by a seeker, which can be either passive or active, to locate and track the target. Once the target is acquired, the weapon is then directed to impact it with high accuracy. Examples of homing munitions include air-to-air missiles, anti-ship missiles, and anti-tank guided missiles.

Loitering munitions, on the other hand, are designed to loiter over an area and wait for a target to appear. These weapons are sometimes referred to as "suicide drones" because they are often designed to crash into the target rather than use an explosive warhead. Unlike homing munitions, loitering munitions are often able to conduct multiple attacks on different targets during a single mission. The Israeli IAI Harop is an example of a loitering munition, which is designed to loiter over an area until it detects an enemy radar system, at which point it can dive to strike the target.

The development and deployment of homing and loitering munitions has become a significant focus for many great powers, as these weapons are seen as effective and efficient ways to conduct military operations. Israel, for example, has developed a range of homing and loitering munitions, including the IAI Harop and the IAI Harpy. These weapons have been sold to a number of countries, including India, Turkey, and China.

In the United States, homing munitions such as air-to-air missiles and anti-tank guided missiles have been a mainstay of military operations for many years. However, the development of loitering munitions has been more recent, with weapons like the Switchblade and the Lethal Miniature Aerial Missile System (LMAMS) gaining increasing attention from military planners.

Russia and China are also developing their own homing and loitering munitions, as part of a wider effort to modernize their military capabilities. Russia's KAB-250 is an example of a homing munition, while China's CH-4 Rainbow is a loitering munition that was built by reverse-engineering the design of the Israeli IAI Harop.

The use of homing and loitering munitions has not been without controversy, as these weapons can sometimes result in collateral damage and civilian casualties. However, their effectiveness in achieving military objectives has led to their increasing use in modern warfare. As the technology continues to evolve, it is likely that homing and loitering munitions will continue to play an important role in military operations.

Unit II. The Evolution of Warfare Technology

Historical Development of Algorithmic Warfare

THE HISTORY OF ALGORITHMIC warfare can be traced back to the earliest use of machines in military operations. The development of early computing technologies in the mid-20th century marked the beginning of a new era of warfare, characterized by increased automation and the use of algorithms in military decision-making.

During the Cold War, both the United States and the Soviet Union developed early forms of algorithmic warfare, such as the use of computers to control missile systems and improve situational awareness. The development of these technologies continued throughout the late 20th and early 21st centuries, with a growing focus on the use of algorithms in military operations.

The first significant use of algorithms in military operations was during the 1991 Gulf War, when the US military used computer systems to control and coordinate air operations. The widespread use of unmanned aerial vehicles (UAVs) in the conflicts in Afghanistan and Iraq marked a new phase in the development of algorithmic warfare, with algorithms being used to control and operate unmanned systems in combat.

The development of artificial intelligence and machine learning technologies in the 21st century has accelerated the integration of algorithms into military operations. Today, algorithms are being used to improve situational awareness, automate routine tasks, and enhance decision-making processes. Autonomous weapons systems, such as drones and autonomous vehicles, are increasingly being developed and deployed, marking a new era in the history of algorithmic warfare. Plans are in place to integrate AI into every aspect of modern military's arsenal, from fighter jets to submarines to missiles, so that they can strike at enemy forces without any human involvement at all.

Major Advances in Algorithmic Warfare

ALGORITHMIC WARFARE has advanced rapidly in recent years, driven by the development and integration of artificial intelligence and machine learning technologies. The integration of artificial intelligence and machine learning technologies into military operations will fundamentally transform the way wars are fought and won. As these technologies continue to evolve, it will be essential to address the ethical, legal, and strategic implications of their use, ensuring that they are used in a responsible and accountable manner.

Some of the key advances in this field include:

- Improved Situational Awareness: Algorithms are increasingly being used to improve situational awareness in military operations, enabling decision-makers to rapidly and accurately assess the operational environment. For example, algorithms can be used to analyze vast amounts of data from sensors and other sources, providing real-time information about the location and behavior of enemy forces.

- Automated Decision-Making: Algorithms are being used to automate routine tasks and enhance decision-making processes in military operations. This includes the use of algorithms to evaluate potential targets, predict enemy movements, and recommend courses of action.

- Autonomous Weapons Systems: One of the most significant advances in algorithmic warfare has been the development and deployment of autonomous weapons systems, including drones and autonomous vehicles. These systems are capable of operating without human intervention, allowing for increased speed and efficiency in military operations.

- Advanced Analytics: Algorithms are being used to process and analyze large amounts of data, allowing military planners to make more informed decisions. For example, algorithms can be used to analyze satellite imagery, social media posts, and other sources of data to identify patterns and trends that can inform military operations.

- Cyber Operations: Algorithms are also being used in cyber operations, including the use of artificial intelligence and machine learning technologies to defend against cyber-attacks and enhance the effectiveness of cyber operations.

These advances in algorithmic warfare have the potential to revolutionize the way wars are fought, offering new operational capabilities and opportunities for military planners. At the same time, they raise significant ethical, legal, and strategic dilemmas, which must be carefully considered as these technologies continue to evolve.

Milestones in Autonomous Warfare Development

THE FOLLOWING TIMELINE highlights some of the key milestones in the history of algorithmic warfare and development of autonomous weapon systems, demonstrating the continued integration of technology into military operations and the growing importance of these technologies in the conduct of warfare. As these technologies continue to evolve, it will be essential to address the ethical, legal, and strategic implications of their use, ensuring that they are used in a responsible and accountable manner.

TIMELINE OF MAJOR MILESTONES in the History of Algorithmic Warfare and Autonomous Weapons:

1940s-1950s:

- Development of early computing technologies and the first electronic computers, such as the ENIAC

and UNIVAC.
- First use of computers in military operations, including the use of computers to control missile systems and improve situational awareness during the Cold War.

1980s-1990s:

- Continued development and integration of computers and algorithms into military operations.
- First use of algorithms in military decision-making during the 1991 Gulf War.

2000s:

- Increased use of unmanned aerial vehicles (UAVs) in military operations, marking a new phase in the development of algorithmic warfare.
- First use of algorithms to control and operate unmanned systems in combat.

2010s:

- Development and deployment of autonomous weapons systems, including drones and autonomous vehicles.
- Increased use of algorithms to improve situational awareness, automate routine tasks, and enhance decision-making processes in military operations.

2020s:

- Continued advancement and integration of artificial intelligence and machine learning technologies into military operations.
- Emergence of new ethical, legal, and strategic dilemmas surrounding the use of algorithmic warfare and autonomous weapons.

Unit III. The State of Algorithmic Warfare Today

Overview of Current Capabilities

ALGORITHMIC WARFARE and autonomous weapons systems have the potential to significantly enhance the capabilities of military forces. These technologies can improve situational awareness, automate decision-making processes, and provide increased speed and efficiency in military operations.

Algorithmic warfare is having an enormous impact on the field of situational awareness. Algorithms are being used to process and analyze large amounts of data from a variety of sources, including sensors, satellites, and social media posts. This information is then used to create real-time situational awareness, allowing military decision-makers to quickly and accurately assess the operational environment.

Autonomous weapons systems, such as drones and autonomous vehicles, are another major area of development in algorithmic warfare. These autonomous systems are capable of operating without human intervention, providing increased speed and efficiency in military operations. Additionally, they offer new operational capabilities, such as the ability to perform reconnaissance and strike missions in difficult or high-risk environments.

Decision-making in military operations is also being improved through the use of algorithms. For instance, algorithms can be used to evaluate potential targets, anticipating enemy movements, and suggesting strategies. This can provide military decision-makers with more accurate and actionable information, enabling them to make better informed decisions.

In the realm of cyber operations, algorithms are being used to defend against cyber-attacks and enhance the effectiveness of offensive cyber operations. This includes the use of artificial intelligence and machine learning technologies to identify and respond to cyber threats, as well as to develop new cyber capabilities and techniques.

Algorithmic warfare and autonomous weapons systems are rapidly advancing and offer new operational capabilities for military forces. However, it is essential to ensure that these technologies are used in a responsible and accountable manner, taking into account the ethical, legal, and strategic implications of their use. The future of algorithmic warfare will continue to evolve and offer new opportunities and challenges for military forces, and it is important to stay informed and prepared as these technologies continue to evolve.

Autonomous Weapons in Contemporary Conflicts

AUTONOMOUS WEAPONS systems have become increasingly prevalent in contemporary conflicts, offering new operational capabilities and opportunities for military forces. However, the use of these technologies in conflict also raises a number of important ethical, legal, and strategic considerations.

One of the key advantages of autonomous weapons systems is their ability to operate in high-risk environments, providing military forces with increased operational flexibility and reducing the risk to personnel. For example, drones can be used to perform reconnaissance missions in areas that are too dangerous for human soldiers, while autonomous vehicles can be used to transport supplies and equipment.

Another advantage of autonomous weapon systems is their speed and efficiency in executing military operations. These systems are able to process and analyze large amounts of data, and make decisions quickly and accurately, without being impacted by fatigue or emotions. This can provide military forces with a decisive advantage in fast-moving or complex operational environments.

However, the use of autonomous weapons systems in conflict also raises a number of important ethical, legal, and strategic considerations. For example, there are concerns about the accountability and responsibility of autonomous weapons systems in the event of unintended harm or collateral damage. There are also concerns about the potential for these systems to be hacked or used for malicious purposes.

In order to ensure the responsible and ethical use of autonomous weapons systems, it is important for military forces to develop clear policies and guidelines for their deployment and use. This should include a comprehensive risk assessment process, which takes into account the potential consequences of their use, as well as the ethical and legal implications.

It is also important for military forces to engage with the international community, including the private sector and academia, in order to better understand the potential consequences and implications of the use of autonomous weapons systems. This will help to ensure that these technologies are used in a responsible and accountable manner, and that the international community is prepared for the challenges and opportunities that lie ahead.

Autonomous weapons systems have the potential to offer significant operational advantages for military forces in contemporary conflicts. However, it is essential to ensure that their use is guided by clear policies and guidelines, and that their potential implications are thoroughly understood and considered. As these technologies continue to evolve and play an increasingly important role in conflict, it is important to stay informed and prepared, and to ensure that they are used in a responsible and ethical manner.

Trends and Predictions for the Future

THE USE OF AUTONOMOUS weapons systems is expected to continue to grow in the coming years, with advancements in technology and increasing demand from military organizations.

One trend that is expected to emerge is the development of more advanced and sophisticated autonomous weapons systems, with greater capabilities in areas such as intelligence gathering, decision-making, and target identification. This will likely result in increased operational flexibility for military forces and greater efficiency in the execution of military operations.

Another trend is the increasing integration of autonomous weapons systems into larger military networks and systems. This will allow military organizations to more effectively leverage the capabilities of these systems, and to coordinate and manage their use more effectively.

Additionally, there is a growing interest in the development of autonomous weapons systems for defensive purposes, such as air and missile defense systems, and for ensuring the security of critical infrastructure and national borders.

As the use of autonomous weapons systems continues to evolve, it is likely that new challenges and opportunities will arise. It will be important for military organizations, governments, and other stakeholders to carefully consider the implications of these trends and to develop strategies for the responsible and ethical use of autonomous weapons systems.

Algorithmic warfare and autonomous weapons systems will continue to shape the nature of conflict and military operations in significant ways. It is crucial that we understand these trends and predictions, and that we work to ensure that these systems are used in a responsible and ethical manner.

Unit IV. Waging Algorithmic Warfare

Operationalizing Artificial Intelligence for Algorithmic Warfare

ARTIFICIAL INTELLIGENCE (AI) has the potential to revolutionize the way that wars are fought and is already being used in a number of innovative ways to enhance the capabilities of military forces. In order to effectively operationalize AI for algorithmic warfare, however, it is necessary to consider a number of key factors.

First and foremost, it is important to have a clear understanding of the goals and objectives of using AI in warfare. This requires a comprehensive understanding of the military requirements, as well as the current state of AI technology and its capabilities.

Once these objectives have been identified, the next step is to develop a robust and scalable AI architecture that can be integrated into existing military systems and processes. This requires close collaboration between military planners, AI engineers, and data scientists, as well as the development of appropriate data infrastructure and algorithms.

Another important consideration is the development of an effective training and simulation environment for AI systems. This will help to ensure that the systems are able to operate effectively in real-world scenarios and can be rapidly updated and adapted to changing circumstances.

Lastly, it is essential to establish a robust governance framework that ensures the responsible and ethical use of AI systems. This includes the creation of clear policies and procedures for the deployment and use of AI, as well as the implementation of appropriate mechanisms for oversight and accountability.

Operationalizing AI for algorithmic warfare requires a comprehensive strategy that takes into account military requirements, the current state of technology, the development of appropriate architectures, effective training and simulation environments, and robust governance. By considering these factors, military organizations can harness the power of artificial intelligence to enhance their capabilities and better prepare for future challenges.

Lethal Autonomous Weapons Systems (LAWS)

Lethal Autonomous Weapons Systems (LAWS) have been a hot topic of discussion in the field of algorithmic warfare. LAWS refers to weapons systems that have the capability to locate, identify, and engage targets without human intervention. In effect, LAWS are fully autonomous "human out of the loop" weapons that operate without direct human supervision.

It is undeniable that LAWS will revolutionize warfare by making it faster, more efficient, and drastically reducing casualties among friendly forces, despite the fact that their development and use raise significant ethical and legal concerns.

One of the biggest debates around LAWS revolves around their compliance with international humanitarian law (IHL). IHL, also known as the laws of war, governs the conduct of armed conflicts and seeks to protect civilians and civilian objects from the effects of hostilities. Some argue that LAWS may violate IHL by being unable to distinguish between military targets and civilians, leading to indiscriminate attacks.

The Department of Defense Directive (DODD) 3000.09 requires that all LAWS be designed to "allow commanders and operators to exercise appropriate levels of human judgment over the use of force" and that "adequate training, [tactics, techniques, and procedures], and doctrine are available, periodically reviewed, and used by system operators and commanders to understand the

functioning, capabilities, and limitations of the system's autonomy in realistic operational conditions."

Additionally, the DODD 3000.09 directive requires that the software and hardware of all Lethal Autonomous Weapons Systems be tested and evaluated to ensure they "function as anticipated in realistic operational environments against adaptive adversaries; complete engagements in a timeframe consistent with commander and operator intentions and, if unable to do so, terminate engagements or seek additional human operator input before continuing the engagement; and are sufficiently robust to minimize failures that could lead to unintended engagements or to loss of control of the system to unauthorized parties."

While military and defense experts consider LAWS to be a necessary step in the evolution of warfare, many are concerned that the development and use of LAWS could lead to an arms race and increase the likelihood of conflict. As with any new weapons system, countries are likely to develop and deploy LAWS to maintain a military advantage. This could lead to a situation where countries are more willing to engage in conflict due to the perceived reduced risk to their own forces.

Letha Autonomous Weapons Systems are already in use in a limited capacity and are likely to become more prevalent in the future. According to a recent United Nations Security Council report on the civil war in Libya, a Lethal Autonomous Weapons Systems (LAWS) may have killed humans for the first time in the spring of 2020. The United Nations report claims that drones made in Turkey called STM Kargu-2 were used to launch an airstrike against the Libyan National Army on behalf of the Libyan Government of National Accord. This attack was carried out by weapons systems with no known human operators "in the loop." According to the UN report, this is the first known instance of the use of a Lethal Autonomous Weapons Systems (LAWS) to kill a human. This even illustrates the reason why it is important that to continue discussions around LAWS and that a framework is established to ensure their responsible and ethical use.

The UN report claims that: "Logistics convoys and retreating [Haftar Affiliated Forces] were subsequently hunted down and remotely engaged by the unmanned combat aerial vehicles or the lethal autonomous weapons systems such as the STM Kargu-2 and other loitering munitions. The lethal autonomous weapons systems were programmed to attack targets without requiring data connectivity between the operator and the munition."

Annex 30 Infographic for STM Kargu-2 loitering munition

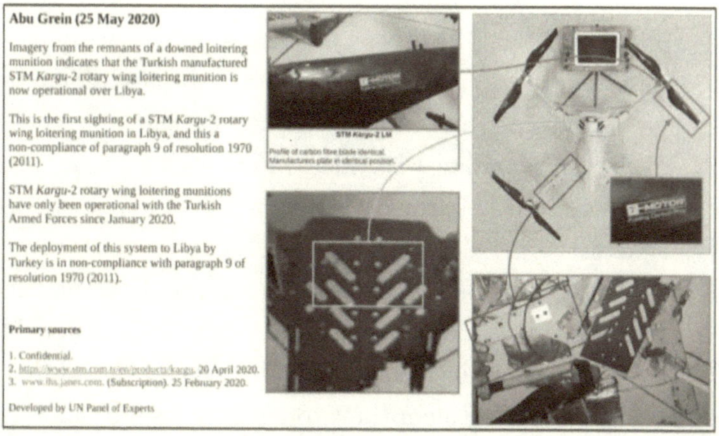

THE STM KARGU-2 IS a mass-produced flying quadcopter that weighs only 7 kg, is capable of fully autonomous targeting, can gather in swarms, continues to function normally even when GPS and radio links are jammed, and can identify and track individual humans using facial recognition software

As LAWS continue to evolve, it is crucial that the international community works together to establish clear guidelines and regulations. This will ensure that LAWS are used in a manner that is consistent with international law and that the risks associated with their use are effectively managed. Failure to do so could have far-reaching and potentially disastrous consequences

Development of Autonomous Weapons by

Global Powers

THE DEVELOPMENT OF autonomous weapons is a rapidly growing field that has attracted the attention of major global powers. The United States, Russia, and China are some of the largest military spenders in the world, and they are all investing heavily in the development of autonomous weapons systems.

The US has been at the forefront of this development with its Project Maven initiative. Launched in 2017, Project Maven is a Department of Defense (DoD) program that aims to use machine learning algorithms to enhance military intelligence and improve decision-making in the field. The project has received a significant amount of funding and has been met with both criticism and support. Critics argue that the use of autonomous weapons could lead to unintended consequences and could increase the risk of collateral damage. Supporters argue that these weapons will increase the efficiency of military operations and reduce the risk to human soldiers.

Russia and China are also investing heavily in the development of autonomous weapons. Russia has developed a range of unmanned systems, including unmanned aerial vehicles (UAVs), unmanned ground vehicles (UGVs), and unmanned underwater vehicles (UUVs). China has been working on autonomous weapons for a number of years and has developed a range of unmanned systems for both military and civilian purposes.

While the development of autonomous weapons is a global trend, each country approaches it differently. The US, for example, has emphasized the importance of ethical considerations and the need to ensure that these weapons are used responsibly. Russia and China, on the other hand, have been less forthcoming about their intentions and have not been as transparent about their development programs.

The development of autonomous weapons by global powers raises important questions about the future of warfare. As these weapons become more advanced, they will have the potential to change the nature of conflict, and it is important that we carefully consider the implications of this technology. The international community must work together to establish clear guidelines and regulations for the development and use of autonomous weapons, to ensure that they are used in a responsible and ethical manner.

Third Offset Strategy (US DOD)

The Third Offset Strategy is the current long-term and strategic plan of the United States to maintain defensive and offensive superiority. It is a set of initiatives taken by the United States Department of Defense to address the challenges posed by technological advancements in areas like artificial intelligence and autonomous systems. The strategic plan was first introduced in 2014, and its primary objective is to maintain the military advantage of the United States in the face of rapidly evolving technological developments.

The Third Offset Strategy involves combining the efforts of humans and machines to boost performance in all areas, including decision-making, precision, productivity, and mechanization. This collaboration, which is increasingly involving machine learning and artificial intelligence, is in response to current and potential future conflict with major powers.

As technology continues to advance and new threats emerge, the ability of the U.S. military to effectively respond and adapt to changing circumstances will be critical to maintaining its strategic advantage in the years to come. As Deputy Defense Secretary Robert Work put it: "We have a resurgent Russia and a rising China. We need to focus on these high-end adversaries."

The Third Offset strategy seeks to integrate new technologies and capabilities into the U.S. military, with a particular focus on areas such as autonomous systems, hypersonic, and biotechnology. The goal is to create a new generation of military systems that can provide enhanced capabilities and greater strategic flexibility.

The most remarkable aspect of the Third Offset is the manner in which the government is pursuing this strategy. When it comes to private sector opportunities, the sky is the limit for small entrepreneurs and innovators whose disruptive thinking is developing new technology to place on the battlefield (the wearable technology suit used by special mission units is one example) or finding new applications for existing technology (weaponizing small drones and developing a counter-drone program).

Improvements in miniaturization (cameras and sensors) and/or computing power at the intersection of machine learning, artificial intelligence, and autonomy make many of these concepts and prototypes feasible (algorithms to provide facial recognition, overhead image analysis, or pattern of life analysis). In the words of Secretary Work once more: "At its core, AI and autonomy will lead to a new era of human-machine collaboration."

PROJECT MAVEN

Project Maven (also known as the Algorithmic Warfare Cross-Functional Team) is a United States Department of Defense (DoD) initiative aimed at harnessing the power of artificial intelligence (AI) for military applications. Launched in April 2017, Project Maven is focused on enhancing the DOD's ability to process, analyze and exploit full-motion video from unmanned aerial vehicles (UAVs). The program is an important step in the Department's broader effort to take advantage of rapidly evolving commercial AI technologies and integrate them into its operations and processes.

Project Maven has been a major driver of AI development within the DoD and has played a key role in advancing the military's use of autonomous systems. The project's success has helped to establish the Department as a leader in the development and deployment of AI-powered technologies and has helped to spur the development of new and innovative applications of AI within the military.

One of primary objective of Project Maven is to rapidly deploy systems capable of applying artificial intelligence, machine learning and algorithmic solutions to sorting and analyzing the vast amounts of intelligence data collected by Predator drones. According to the Department of Defense, Project Maven improves the performance of human-machine teams by combining intelligence and operations through AI/ML and augmented reality technology. Project Maven has enabled the US military to reduce decision-making time to a fraction of what it would be without AI/ML.

One of the key challenges that Project Maven is trying to address is the vast amount of video footage captured by UAVs, which can be difficult to process and analyze in a timely manner. Project Maven is working to develop algorithms that can automatically extract and analyze meaningful information from this video data, which can then be used to support military operations, enhance situational awareness, and improve decision-making.

Project Maven focuses on analyzing full-motion video data collected by tactical aerial drone platforms like the ScanEagle and medium-altitude platforms like the MQ-1C Gray Eagle and the MQ-9 Reaper. Each day, a single drone equipped with these sensors generates a huge amount of data, often in the terabyte range. Before artificial intelligence (AI) was applied, it would have taken a team of analysts 24 hours a day to exploit just a small portion of the sensor data collected by a single drone.

The Department has been working closely with a number of commercial technology companies and academic institutions to develop and refine these algorithms, and the results have been promising. Project Maven has already delivered several AI-powered capabilities that are now being used by military personnel in the field, including object recognition, target tracking and event detection.

As a major investment in the research, development, and eventual deployment of autonomous systems, Project Maven is a crucial component of the military's effort to reap the benefits of AI. Project Maven, by focusing on AI-powered capability development, is assisting in laying the groundwork for the military's future use of autonomous systems and ensuring that the Department stays at the forefront of this rapidly evolving field.

Robert Work, who was then Deputy Secretary at the US Department of Defense (DOD), created the Algorithmic Warfare Cross-Functional Team (or "Project Maven") in late April 2017 with a memo. He wrote "As numerous studies have made clear the department of defense must integrate artificial intelligence and machine learning more effectively across operations to maintain advantages over increasingly capable adversaries and competitors."

"Although we have taken tentative steps to explore the potential of artificial intelligence, big data and deep learning," he added, "I remain convinced that we need to do much more and move much faster across DoD to take advantage of recent and future advances in these critical areas."

At first, there were only six people on the team. Lt. Gen. Jack Shanahan of the Air Force, and Col. Drew Cukor of the Marine Corps, were overseeing the development and deployment of Project Maven, which was at that time the first operational use of deep learning AI technologies within the defense intelligence enterprise.

Semi-autonomous systems have been in use for some time, and the US Department of Defense (DOD) has long supported AI's early stages of development through funding for basic research and development. However, the DOD had never before attempted to use deep learning and neural networks on par with cutting-edge commercial AI in departmental operations in a combat theater as they have with Project Maven.

Experts in the field of artificial intelligence from both the private and public sectors advised the Department of Defense against launching Project Maven until they found a data-intensive problem with limited human casualties. Fortunately for the Project Maven team, the defense intelligence community was at the time inundated with data gathered by US drones, spy planes and satellites which was more than the Department of Defense could ever hope to analyze in a lifetime of full-time work.

Despite being tasked with leading an artificial intelligence project, none of the six original members of Project Maven had any prior experience with either field. Instead, they had to focus on forming alliances, first with AI specialists in the private sector and academic institutions, and then with drone sensor analysts working for the Pentagon.

Using resources provided by Defense Information Unit Experimental, a group formed to accelerate the Department of Defense's adoption of commercial technologies, the Project Maven team was able to enlist the aid of prominent experts in artificial intelligence. Simply figuring out how to involve the tech sector in projects successfully is a huge accomplishment.

Project Maven was able to start off on the right foot by assembling a strong team and forming strategic alliances. Before Maven, no one in the Department of Defense knew how to purchase, deploy, or implement artificial intelligence. The standard defense acquisition process takes several years and involves multiple departments and agencies working in isolation to define the roles that acquisitions must play and develop, manufacture, and deploy the necessary technologies. With digital technologies, this strategy often ends up producing underperforming systems that become obsolete before they are even put into production.

The Project Maven team took a new approach, one inspired by commercial technology project management practices. Prototypes of the product and the system that supports it are built iteratively and constantly refined based on feedback from the user base. Users can help developers better meet their needs, and users can help their organizations be ready to quickly and effectively implement AI capabilities. Labeling data, creating AI-computational infrastructure, creating, and integrating neural net algorithms, and gathering user feedback are all crucial activities that run in parallel and iteratively throughout the development of an AI system.

Despite the impressive capabilities of today's AI techniques for image analysis, it is still not simple or "plug-and-play" to create specialized algorithms for a given application. In order to train the deep learning algorithm, which is essential for creating robust AI systems, enormous data sets are required. Not only must training data be accessible, but it must also be manually labeled and organized. This stage of automation is paradoxically very labor-intensive. In the case of Maven, the first training data sets required the manual labeling of more than 150,000 images by humans. Robust performance across the vast range of possible operating conditions, including variations in altitude, density of tracked objects, image resolution, view angle, and so on, necessitates the use of such large training data sets.

Once the labeled data is prepared, the AI model training process requires substantial computational power. For such calculations, conventional information technology infrastructure is essentially useless. A large number of the prominent businesses in the tech industry have gone so far as to design and implement their own custom processors and cloud infrastructure networks specifically for use in AI computations. The Defense Department had already spent billions of dollars and years trying to move its digital operations to the cloud, but none of the existing infrastructure met the needs of AI testing and training. Project Maven had to construct its own AI-ready infrastructure from scratch, including computing clusters for graphics processing.

Defense Innovation Unit

The Defense Innovation Unit (DIU) is a United States Department of Defense organization that was established in 2015 to accelerate the adoption of emerging technologies into the military. The DIU is focused on identifying and testing commercial technologies with potential military applications, as well as developing new technology solutions to address specific military needs. The DIU has been particularly active in the development and adoption of autonomous systems, including autonomous weapons.

The DIU, which is headquartered in Mountain View, California with offices in Boston, Austin, Chicago, and the Pentagon just outside Washington, D.C., manages six portfolios focused on meeting national security needs in the areas of artificial intelligence (AI), autonomous systems (AS), cyberspace (CYBER), human systems (HUMAN), energy (ENERGY), and space (SPACE).

One of the key goals of the DIU is to leverage the capabilities of the private sector to develop new technology solutions for the military. To achieve this goal, the DIU works closely with technology companies, startups, and other organizations to identify and evaluate new technologies that could be adapted for military use. The DIU provides funding, technical expertise, and other resources to support the development and testing of these technologies.

One of the key areas of focus for the DIU is the development of autonomous systems, including autonomous weapons. The DIU has played a significant role in the development and adoption of a number of autonomous weapons systems, including unmanned aerial vehicles (UAVs), unmanned ground vehicles (UGVs), and autonomous maritime systems.

The DIU has been particularly active in the development of UAVs, which are increasingly being used for a wide range of military applications, including surveillance, reconnaissance, and targeted strikes. The DIU has been working with a number of private sector companies to develop new UAV technologies that are better suited to military use, including longer endurance, higher altitude, and increased payload capacity.

In addition to developing new technologies, the DIU is also focused on accelerating the adoption of existing technologies into the military. To achieve this goal, the DIU works closely with military organizations to identify areas where new technology solutions can be applied, and then helps to integrate these solutions into existing military systems.

The DIU has been praised for its innovative approach to technology development and its ability to rapidly prototype and test new solutions. However, the organization has also faced some criticism for its close ties to the private sector, which some argue could lead to conflicts of interest and other ethical concerns.

The Defense Innovation Unit has played a significant role in the development and adoption of autonomous weapons and is likely to continue to be a major player in this field in the years to come. While the organization has faced some criticism and concerns over its close ties to the private sector, its ability to rapidly develop and test new technologies has been a significant asset to the military.

Unit V. Risks and Benefits of Autonomous Weapons Systems

Potential Risks of Algorithmic Warfare

AS THE USE OF AUTONOMOUS weapons systems continues to propagate, it is important to consider the potential risks and unintended consequences associated with their use.

Among the risks associated with the use of autonomous weapons systems is the potential for unintended harm, collateral damage, and misidentification. For instance, will autonomous weapons be able to tell the difference between enemy troops and kids with toy guns when deciding whom to shoot? Between civilians trying to get away from the fighting and insurgents retreating for strategic purposes?

Paul Scharre, an expert on autonomous weapon systems, uses the analogy of the runaway gun to explain the issue with using autonomous weapons. As the name implies, a runaway gun is a malfunctioning machine gun that keeps firing even after the trigger has been released. Because the gun "does not know it is making a misidentification error," it will keep firing until all of the ammunition is gone. Extremely dangerous runaway guns have human operators who can cut the ammunition link or try to reorient the weapon away from potential danger. The very nature of autonomous weapons precludes any such safeguard from being present.

Nevertheless, we can mitigate these risks by having robust systems in place for monitoring and controlling the use of autonomous weapon systems, including measures to ensure that they comply with international humanitarian law and other relevant ethical and legal frameworks.

Another major risk associated with the use of autonomous weapons systems is the potential for these systems to be hacked or used for malicious purposes. To mitigate this risk, it is important to have strong cybersecurity measures in place, and to ensure that these systems are designed and implemented with robust security protocols.

There is a possibility that the use of autonomous weapons systems could result in an increase in the number of conflicts, or in conflicts that are more intense and more destructive. To mitigate this risk, it is essential to engage in diplomatic efforts to reduce tensions and promote stability and peace, as well as to ensure that the use of these systems is guided by a set of clear and transparent ethical and legal principles. It is essential to consider the long-term repercussions of the use of autonomous weapons systems, including the impact on international security, the balance of power, and the development of new technologies and capabilities.

The potential uncontrolled spread of autonomous weapons is another major threat brought on by algorithmic warfare. Economies of scale and consumer demand could lead to the production and distribution of autonomous robots that are low-cost, highly lethal, and difficult to stop once they've spread around the world. Those who shouldn't have access to these weapons, like non-state actors, organized crime groups, foreign and domestic terrorists, could end up with "Kalashnikov" autonomous weapons.

Because it essentially eliminates two of the major concerns that have historically prevented and shortened wars—concern for civilians abroad and concern for one's own soldiers—autonomous weapon system are likely lead to more frequent wars. At the same time, because autonomous weapon system will significantly reduce if not altogether eliminate the need for nations to risk the lives of human soldiers, their use will significantly alter the cost-benefit analysis that nations perform prior to initiating and launching wars.

Wars fought on the territory of countries with inferior technology, or "asymmetric wars," are expected to become much more frequent with the advent of autonomous weapon systems. Consider the ripple effects felt all over the world as a result of the Cold War's military interventions by the Soviet Union and the United States, from the very first proxy war to the present day. If you then extrapolate that to every country that is developing autonomous weapons of any kind, the result is staggering.

It's also likely that powerful rich countries will race to build the deadliest autonomous weapons possible, including those that can mount CBRN (chemical, biological, radiological, and nuclear) weapons. Expanding the use of these weapons would amplify the moral risks posed by their increasing lethality.

Autonomous Weapons Capable of Launching Chemical, Biological, Radiological, and Nuclear (CBRN) Attacks

Military leaders and policymakers are becoming increasingly concerned about the development of autonomous weapons capable of launching CBRN attacks. Chemical, biological, radiological, and nuclear weapons can have a devastating impact on both military and civilian populations, and the use of autonomous systems in such attacks could lead to even greater destruction.

The lack of human oversight is one of the biggest concerns with autonomous weapons that can launch CBRN attacks. There is a risk that autonomous weapon systems may not be able to adapt to changing battlefield conditions, and that they will instead pursue objectives that are inconsistent with the goals of the mission.

For example, an autonomous system may be programmed to launch a CBRN attack in response to certain conditions but may not have the ability to distinguish between military targets and civilian populations. This could lead to significant civilian casualties and significant political fallout.

Another major concern is the potential for malicious actors to use autonomous weapons to mount CBRN attacks. Terrorist organizations, for example, may seek to use autonomous systems to spread disease or to release toxic chemicals into populated areas. In the wrong hands, these weapons could be used to cause widespread destruction and to disrupt critical infrastructure, potentially leading to a global catastrophe.

However, the benefits of developing autonomous weapons that can carry out CBRN attacks outweigh the risks. These autonomous systems may be used to respond rapidly to CBRN threats, decreasing the severity of an attack or preventing one altogether. Furthermore, autonomous weapon systems may be able to operate in conditions that are too hazardous for human soldiers, potentially reducing the risk of casualties among military personnel.

The emergence of autonomous weapons with the potential to launch CBRN attacks is a difficult and nuanced topic that calls for in-depth research and analysis. Policymakers and military leaders must evaluate the potential risks and benefits of these systems and collaborate to develop a framework for the ethical and responsible use of autonomous weapons. This will require close cooperation among the military, intelligence, and scientific communities, as well as between nations and international organizations.

AI in Nuclear Weapon Systems

THERE HAS BEEN GROWING concern in recent years about the integration of artificial intelligence and machine learning in nuclear weapons systems. While the deployment of artificial intelligence into nuclear weapon systems has the potential to significantly improve response speed and accuracy, it also raises serious concerns about risk of accidental use and the delegation of human decision-making to machines.

On one hand, AI-powered nuclear defense systems could potentially lessen the time required to detect and respond to a nuclear threat, thereby enhancing the efficacy and reliability of deterrent strategies and reducing the risk of accidental use. For instance, AI algorithms could be used to analyze satellite imagery, radar, and other data sources in real time to detect ballistic missile launches. This data could then be used to guide the deployment of anti-missile systems and assist decision-makers in responding swiftly and effectively to a potential attack.

On the other hand, the integration of AI in nuclear weapons systems raises significant concerns about the possibility that the speed and precision of AI-powered decision-making could escalate conflicts, leading to rapid and unintentional escalation to the use of nuclear weapons. Moreover, the delegation of decision-making from humans to machines in nuclear weapon systems is highly controversial, with some experts raising questions about the ethical implications of machines making decisions about the use of nuclear weapons.

There is also a growing concern about the risk of accidental use of AI in nuclear weapon systems and the potential for malicious actors to compromise or manipulate AI-powered systems, either through hacking or by feeding false data into the system, further exacerbates the risk of accidental use. As

AI algorithms become more sophisticated, the likelihood of unintended or accidental launches increases, especially given the complex and unpredictable nature of the international security environment.

The integration of artificial intelligence into nuclear weapons is not a new development as artificial intelligence (AI) has long been a staple of nuclear deterrence systems. The United States and the Soviet Union realized in the 1960s that artificial intelligence (AI) could help them maintain their retaliatory capability, or their ability to respond to a surprise nuclear attack. In order to increase the agility of their command-and-control system and give decision-makers more time to concentrate on what really mattered—deciding whether to launch a nuclear strike or not—they pursued the development of AI systems. Artificial intelligence (AI) was first used to automate threat detection, launch order transmission logistics, and missile targeting and guidance.

Nuclear-armed states recognized the potential of AI for nuclear deterrence early on, but they also recognized its shortcomings. They were hesitant to delegate the decision making to AI systems due to the dire repercussions that an accidental launch would have. The Soviet Union was the only nation to actively pursue the creation of a nuclear arsenal with a fully autonomous command and control system. The Perimetr system was designed to be used only in the extremely unlikely event that devastating nuclear attack completely annihilated the Soviet military nuclear command and control.

In March 2018, Russian president Vladimir Putin announced that Russia would revive and modernize the Perimetr system (commonly referred to as "Dead Hand" in the United States) as an AI-enabled system which can detect a nuclear strike against Russia and launch a guaranteed retaliatory nuclear strike even if the lines of communication between its Strategic Missile Forces are totally destroyed.

The newly revamped Perimetr system will not require any human intervention as it will be fully autonomous. The decision of when and where to launch a nuclear attack will be made by a complex artificial intelligence system, the autonomous nuclear command and control, which analyzes different types of data, such as seismic activity, radiation levels, atmospheric pressure, and the volume of chatter on military radio frequencies in addition to constantly monitoring telemetry from the Russian Strategic Missile Force observation posts and early warning system data (EWS).

In order to determine whether or not there was a massive nuclear strike, the autonomous weapon system collects data on seismic disturbances in the same locations and compares it with data from multiple point sources of powerful ionizing and electromagnetic radiation. If the AI system concludes that a massive nuclear strike occurred, Perimetr would launch a retaliatory strike that would bypass Russia's strategic missile missiles command and control center.

In the event that Russian human decisionmakers receive information from the EWS that other countries have launched a nuclear strike against it, they would activate the Perimeter system. The Perimetr system will launch a retaliatory nuclear strike if it doesn't receive a shutdown command from the strategic missile missiles command and control center within a certain amount of time. This removes the risk associated with human error and guarantees a retaliatory strike even if the Russian command-and-control units have been destroyed.

Russian leaders have repeatedly assured their international counterparts that a accidental or unauthorized nuclear launch is not possible due to the fact that the Perimetr artificial intelligence systems performs four checks just before launching an automatic retaliatory nuclear strike. According to Russian leadership, Perimetr will first determine if a nuclear attack has occurred comparing the data on seismic disturbances to powerful ionizing and electromagnetic radiation data it collects. Then it will check its line of communication with the General Staff of the Russian military. If a communication line is still active, the AI system will deactivate itself. After waiting for a response from the General Staff, Perimetr then contacts Russia's main strategic missiles command and control center. If it is unable to establish a line of communication with either of these bodies, then the artificial intelligence will grant final decision-making authority to the lowest-ranking member of the command bunker. And only then does it begin to act.

The use of autonomous weapons systems presents a number of important risks and challenges, and it is crucial that these risks are carefully considered and effectively managed. By taking a proactive and responsible approach to the development and use of these systems, we can help to ensure that they are used in a manner that supports stability, peace, and security, and that advances our collective interests and values.

Mitigating Risks of Algorithmic Warfare

THE USE OF AUTONOMOUS weapons systems in conflict raises important ethical, legal, and strategic considerations. It is crucial that these risks are mitigated to ensure that these systems are used in a responsible and ethical manner.

To minimize these risks, there are a number of steps that can be taken by military organizations, governments, and other stakeholders. One way to mitigate the risks associated with autonomous weapons systems is to establish clear and transparent governance frameworks. This should include international norms and standards, as well as national policies and regulations, to govern the development, deployment, and use of these systems.

Another important step is to ensure that autonomous weapons systems are subject to rigorous testing and evaluation, to assess their capabilities and limitations, and to identify and address any potential risks or ethical concerns. This can be done through the development of testing protocols, and the involvement of independent experts and stakeholders in the evaluation process.

It is also important to ensure that autonomous weapons systems are designed and operated in a manner that is consistent with the principles of international humanitarian law and the laws of armed conflict. This includes ensuring that these systems are capable of distinguish between military and civilian targets, and that they are able to minimize harm to civilians and civilian objects.

It is important to consider the strategic implications of autonomous weapons systems, and to ensure that their deployment and use is consistent with broader national security and foreign policy objectives. This may include ensuring that these systems are used in a manner that supports stability and security in the international system, and that they do not contribute to an arms race or increase the risk of conflict.

International norms and standards, rigorous testing and evaluation, ethical considerations, and strategic planning are all necessary to reduce the dangers of algorithmic warfare and autonomous weapons systems. Following these procedures will guarantee that these systems are used in an ethical and responsible manner, thereby bolstering global peace and safety.

Benefits of Autonomous Weapons Systems

THE ADVANTAGES OF USING autonomous weapon systems outweigh the associated dangers. In order to make well-informed decisions and efficiently oversee the implementation of these systems, it is essential to recognize these advantages.

The increased efficiency and accuracy of military operations is one of the most important advantages of autonomous weapons systems. By utilizing algorithms and sophisticated sensors, these systems can rapidly identify and engage targets with a high degree of precision, thereby minimizing the risk of unintended harm and collateral damage. This could lead to more effective and accurate military operations, as well as fewer reduced risks for humans.

Autonomous weapon systems have the ability to operate in challenging or high-risk environments, such as the deep ocean, outer space, or areas contaminated by chemical, biological, or radioactive agents. By utilizing these systems, military organizations are able to reduce the risk to human soldiers and gain valuable intelligence and operational capabilities in these difficult environments.

For nuclear-armed states, autonomous weapon systems like drone aircraft or underwater vehicles may represent a viable replacement for ICBMs, manned jet bombers, and submarines. Autonomous version of these weapon system could be deployed for extremely long periods of time (days, months, or even years) without being destroyed (unlike missiles and torpedoes).

One nuclear-armed country is already considering this option: In 2015, Russia admitted it was working on a nuclear-armed, unmanned submarine codenamed Status-6. Later in March 2018, during a speech to the Russian Federal Assembly, Russian President Putin announced that Russia would build a fully autonomous nuclear submarine to be named Poseidon, which has been described by US media as a nuclear-armed, nuclear-powered, autonomous undersea doomsday device.

According to a Kremlin translation of Putin's comments, the autonomous drone would travel quietly to "great depths," and move faster than a submarine or boat, "have almost no vulnerabilities for the enemy to exploit," and "carry massive nuclear ordnance" without making a sound.

Vladimir Putin has said, "Unmanned underwater vehicles can carry conventional or nuclear warheads, allowing them to engage a variety of targets, including aircraft groups, coastal fortifications, and infrastructure."

This new autonomous nuclear doomsday machine is said to be capable of triggering 300-foot tsunamis and has essentially weaponized nuclear radiation itself, which could leave massive swaths of Earth uninhabitable for the better part of a century. This is in contrast to traditional nuclear weapons, which use nuclear detonations in the air to put massive heat and pressure on targets.

The use of autonomous weapons systems has the potential to increase operational speed and responsiveness. These systems can operate more quickly and more independently than human soldiers and can respond to changing circumstances in real-time. This can be particularly important in fast-moving and fluid military situations, such as in the conduct of urban warfare or counter-terrorism operations.

The use of autonomous weapons systems has the potential to reduce the costs associated with the conduct of military operations. By relying on these systems, military organizations can reduce their dependence on human soldiers and associated equipment, potentially leading to reduced costs, increased efficiency, and improved military capability.

There are some significant upsides to employing autonomous weapons systems, despite the risks and difficulties that may be involved. We can help ensure that these systems are used in a way that furthers our collective interests and values by carefully considering the benefits and risks, and by taking a proactive and responsible approach to their development and use.

Balancing Risk and Benefits of Autonomous Warfare

FOR MILITARY ORGANIZATIONS, policymakers, and society at large, the challenge posed by the emergence and utilization of autonomous weapons systems is a daunting and rapidly developing one. One the one hand, there is the possibility that these systems will result in considerable improvements efficiency, precision, and operational capability. On the other hand, they also pose important risks in terms of unintended harm, accountability, and the potential for the erosion of norms and values.

Balancing these risks and benefits is a critical challenge, and requires a careful and informed approach to the development and use of these systems. This can be achieved through a number of key steps, including:

1. Establishing clear policies and guidelines for the development and use of autonomous weapons systems, including guidelines for testing and evaluation, and for ensuring that these systems are used in a manner that is consistent with international law and norms.

1. Developing robust systems of accountability and transparency, including through the establishment of independent oversight mechanisms, the development of standardized reporting systems, and the creation of clear lines of responsibility for the use of these systems.

1. Fostering a robust and open dialogue between stakeholders, including military organizations, policy makers, industry, civil society, and academic experts, in order to better understand the risks and benefits associated with these systems and to identify best practices for their development and use.

1. Encouraging investment in research and development to better understand the capabilities and limitations of these systems, and to help ensure that they are used in a responsible and effective manner.

1. Promoting education and awareness about the challenges and risks associated with the use of autonomous weapons systems and working to engage the public in a broader conversation about the future of warfare and the role of technology in shaping our security and defense.

Ultimately, balancing the risks and benefits of algorithmic warfare requires a proactive, informed, and collaborative approach to the development and use of these systems. By working together, we can help ensure that they are used in a manner that advances our collective interests and values, while mitigating the risks associated with their use.

Unit VI. International Responses to Autonomous Weapons

National and International Regulations

THE USE OF AUTONOMOUS weapons systems raises important legal, ethical, and policy questions, and calls for a clear and comprehensive framework of national and international regulations. To date, there has been limited progress in this area, with a patchwork of national policies and international agreements that offer varying levels of guidance and protection. However, there is growing recognition of the need for a more coordinated and harmonized approach to regulating these systems, in order to ensure that they are used in a manner that is consistent with international law and norms.

There are a number of key considerations when it comes to regulating the use of autonomous weapons systems, including:

1. National policies and regulations: Individual countries have a critical role to play in shaping the development and use of autonomous weapons systems, and many have begun to develop policies and regulations to govern their use. These policies may vary widely in scope and content and may focus on issues such as the development and use of these systems, testing and evaluation, and accountability and transparency.

1. International agreements and norms: The

international community has also played an important role in shaping the development and use of autonomous weapons systems, through a variety of international agreements and norms. These agreements include the Convention on Certain Conventional Weapons (CCW), the Arms Trade Treaty (ATT), and the Rome Statute of the International Criminal Court (ICC), among others.

1. Ethical and moral considerations: In addition to legal and policy considerations, the development and use of autonomous weapons systems also raises important ethical and moral questions, including questions of accountability, responsibility, and the potential for unintended harm. These questions must be carefully considered and addressed as part of any comprehensive regulatory framework.

1. Balancing risks and benefits: Finally, it is important to ensure that any regulatory framework takes into account the potential risks and benefits of these systems, and that regulations are designed to maximize the benefits while mitigating the risks. This requires a nuanced and informed understanding of the capabilities and limitations of these systems, as well as a deep understanding of the broader political and military context in which they are used.

The future of algorithmic warfare will be heavily influenced by national and international regulations that set standards for how these systems can be used in accordance with international law and norms. To ensure that these systems are used in a responsible and effective manner as they continue to evolve and play an increasingly central role in our security and defense, it will be important to keep establishing and refining our regulatory frameworks.

International Cooperation in Algorithmic Warfare

INTERNATIONAL COOPERATION in algorithmic warfare is essential for effectively addressing the complex and rapidly escalating threats posed by autonomous weapons. Collaboration between nations enables a greater understanding of the capabilities and limitations of these weapons, as well as the identification of common goals and the exchange of best practices.

The development of common norms and standards for the use of autonomous weapons is one area where international cooperation can be incredibly valuable. This may involve establishing ethical guidelines for the design and use of these weapons, as well as protocols for the deployment and operation of autonomous systems in the field.

The exchange of information and knowledge regarding the latest advancements in algorithmic warfare is another area where international cooperation can be advantageous. This could involve the exchange of information and insights regarding the use of autonomous weapons in real-world scenarios, as well as best practices for their deployment and use.

International cooperation in algorithmic warfare could also address the difficulty of ensuring the accountability of autonomous weapons. This can be accomplished by establishing common norms and standards for the use of these weapons and by encouraging the sharing of information and data regarding their deployment and use in real-world scenarios.

The Role of International Organizations

INTERNATIONAL ORGANIZATIONS play an essential role in shaping the future of algorithmic warfare by fostering international cooperation and collaboration, establishing norms and standards for the use of autonomous weapons, and addressing the complex challenges posed by these systems.

The United Nations is one of the most important international organizations in this field, and it has been promoting the responsible use of autonomous weapons through a number of initiatives. For instance, the Convention on Certain Conventional Weapons (CCW) has been developing norms and standards for the use of these weapons, while the United Nations Office for Disarmament Affairs (UNODA) has advocated for the development of an international legal framework for the regulation of autonomous weapons.

The International Committee of the Red Cross (ICRC) is another key player in the field of algorithmic warfare, working to promote the responsible use of autonomous weapons in accordance with international humanitarian law. This includes advocating for the development of ethical guidelines for the use of these weapons and promoting the international exchange of information and best practices.

In addition to these organizations, several non-governmental organizations (NGOs) are also active in this area. These organizations work to raise awareness of the challenges posed by autonomous weapons and to promote the ethical and responsible use of these systems.

The future of algorithmic warfare will be shaped significantly by international organizations. By promoting cooperation and collaboration between nations, establishing norms and standards for the use of autonomous weapons, and addressing the complex challenges posed by these systems, these organizations can contribute to the responsible and ethical use of algorithmic warfare.

Unit VII. Ethical and Legal Considerations of Algorithmic Warfare

Debates around Autonomous Weapons and International Law

THE ROLE THAT INTERNATIONAL law plays in regulating the use of autonomous weapons is one of the most contentious issues in the world of algorithmic warfare. Proponents of autonomous weapons argue, on the one hand, that they have the potential to save lives by reducing the number of soldiers required to engage in combat, reducing the risk of human error, and increasing the speed and efficiency of military operations. Opponents, on the other hand, argue that autonomous weapons pose a substantial risk to human life and violate international law by removing human control over the use of force.

International humanitarian law, also known as the law of armed conflict, is the body of law that regulates the conduct of armed conflicts. It governs the methods and means of warfare and seeks to limit the effects of conflict on civilians and civilian objects. The key principles of international humanitarian law include distinction, proportionality, and military necessity. Distinction requires that military operations be directed at military objectives only and that civilians and civilian objects be spared as much as possible. Proportionality requires that the expected civilian harm not be excessive in relation to the anticipated military advantage. Military necessity requires that the use of force be proportional and limited to what is necessary to achieve a legitimate military objective.

Autonomous weapons raise questions about the application of these principles to autonomous systems. For example, can autonomous weapons distinguish between military and civilian targets, and if so, how accurate are they in doing so? Can they assess the proportionality of the use of force and comply with the principle of military necessity? These are critical questions that must be addressed in order to ensure that autonomous weapons are used in a manner that is consistent with international law.

International organizations and human rights groups have called for a ban on autonomous weapons, citing concerns about accountability, transparency, and the potential for abuse. They argue that autonomous weapons raise serious ethical and moral concerns and that the use of such weapons could lead to a future in which wars are fought without human intervention. Some countries have also raised concerns about the development of autonomous weapons, particularly in light of recent advancements in artificial intelligence and robotics technology.

There is no clear consensus on how to regulate the use of autonomous weapons, and the debate surrounding them, and international law is ongoing. As technology continues to advance and the use of autonomous weapons becomes more prevalent, it is imperative that international actors collaborate to develop clear, effective, and binding regulations to ensure that these weapons are used in accordance with international law and with respect for human life.

The Ethics of Algorithmic Warfare

THE USE OF AUTONOMOUS weapons has sparked a heated ethical debate within the military and international community. Some argue that the use of autonomous weapons is necessary for the safety of military personnel and the success of missions, whereas others believe that the use of these weapons raises serious ethical concerns regarding accountability and the protection of civilians.

The issue of accountability is one of the most significant ethical considerations regarding autonomous weapons. Who is accountable if an autonomous weapon malfunctions or makes a decision that causes civilian casualties? If a weapon is programmed to make decisions based on algorithms, is it ethical to hold the programmer, the military, or the government accountable?

The possibility that these weapons could be used to commit violations of human rights or war crimes is a further ethical concern. The use of autonomous weapons raises questions about whether or not these weapons can be programmed to differentiate between civilian and military targets and adhere to the laws of armed conflict.

In addition to these concerns, there is also the question of whether the use of autonomous weapons is consistent with the principles of just war. Some argue that the deployment of autonomous weapons goes against the principle of discrimination, which requires that military action be directed only at legitimate military targets and that the use of force be proportionate to the threat posed.

There is no simple answer to the complex and nuanced ethical debate surrounding algorithmic warfare. The international community must engage in a robust and ongoing dialogue about the ethical implications of autonomous weapons to ensure that their deployment is guided by a clear and well-informed ethical framework.

The international community must address a number of grave ethical concerns raised by the use of autonomous weapons. The deployment of these weapons must be guided by ethical principles and international regulations that ensure accountability, protect civilians, and uphold the principles of just war, despite the fact that they have the potential to offer significant benefits in terms of military efficiency and safety.

Addressing the Challenges of Algorithmic Warfare

THE CHALLENGES OF ALGORITHMIC warfare are numerous and complex. To address these challenges, a multifaceted approach is required that incorporates both technical and non-technical measures. This section of the book will discuss some of the key challenges associated with algorithmic warfare and explore potential solutions to these challenges.

The possibility of unintended consequences is one of the major obstacles associated with algorithmic warfare. The operation of autonomous weapon systems is based on algorithms and data that are susceptible to biases and errors. This can lead to unintended and harmful consequences, such as the targeting of civilians and noncombatants. To mitigate this risk, rigorous testing and evaluation procedures must be in place to ensure that autonomous weapons systems operate safely and effectively.

The need to ensure accountability and transparency is another difficulty associated with algorithmic warfare. Autonomous weapon systems can be operated by a single person or a small group with minimal oversight. This can lead to decisions that are inconsistent with international laws and standards. It is essential to have clear and transparent mechanisms in place to ensure that the use of autonomous weapons is subject to appropriate oversight and accountability in order to address this challenge.

The potential for proliferation and the risk of destabilization also pose a difficulty for algorithmic warfare. As more nations develop and deploy autonomous weapon systems, the possibility of unintended escalation or accidental conflict rises. To address this issue, it is essential to engage in international dialogue and cooperation to ensure that autonomous weapons systems are utilized responsibly and in a predictable manner.

Finally, it is necessary to consider the ethical implications of algorithmic warfare. The use of autonomous weapons raises a number of complex moral and ethical questions, such as the morality of delegating the decision to kill to machines and the responsibility for actions taken by these systems. Taking into account the perspectives of a variety of stakeholders, such as military leaders, philosophers, and human rights experts, a thoughtful and nuanced approach is required to address these ethical considerations.

To address the complex and multifaceted challenges of algorithmic warfare, a coordinated and all-encompassing strategy that incorporates technical, operational, and ethical considerations is required. It is possible to ensure that algorithmic warfare is used responsibly and sustainably, for the benefit of all, through continued research, development, and engagement.

Unit VIII. The Future of Algorithmic Warfare

Predictions and Scenarios for the Future

AS THE WORLD CONTINUES to develop and integrate autonomous weapons into military operations, it is important to consider what the future may hold. While the future is inherently uncertain, by examining current trends and developments, it is possible to make educated predictions and scenarios for the future of algorithmic warfare.

One likely scenario is that the use of autonomous weapons will become increasingly widespread. With advances in artificial intelligence and machine learning, autonomous weapons are likely to become more capable, versatile, and reliable. This increased capability will make them appealing to a wider range of military users, leading to greater proliferation and use.

Another scenario is that the development of autonomous weapons will drive a new era of strategic competition between major powers. As countries compete to develop and acquire these weapons, it is likely that they will seek to gain strategic advantages over their rivals, which could lead to destabilizing arms races and increased tensions between major powers.

Additionally, it is likely that the international community will face new challenges as it tries to regulate the use of autonomous weapons. As these weapons become more capable and widely used, it will become increasingly difficult to develop and enforce international norms and regulations that effectively address the risks and challenges posed by these weapons.

Finally, it is possible that the development of autonomous weapons will lead to new opportunities for peace and stability. For example, by reducing the risks of human error and friendly fire incidents, autonomous weapons may help to reduce the likelihood of accidents and misunderstandings between military forces. Additionally, by reducing the need for human soldiers to engage in dangerous and traumatic combat operations, autonomous weapons may help to reduce the human costs of war.

While these scenarios represent a range of possible futures, they all underscore the need for continued careful consideration of the risks and benefits of algorithmic warfare. By anticipating the future developments and challenges posed by autonomous weapons, it will be possible to better prepare for the future and make informed decisions about the role these weapons will play in shaping the global security landscape.

The Impact of Autonomous Weapons on Society

IN RECENT YEARS, DISCUSSIONS of autonomous weapons have dominated the headlines, ranging from the benefits they may provide to the ethical and moral questions they raise. As we move towards a future where algorithmic warfare becomes increasingly prevalent, it is important to consider how these advancements will impact society as a whole.

The effect on the economy and the labor force is likely to be one of the most profound as the deployment of autonomous weapons systems will likely lead to a reduction in the number of personnel needed for military operations. This could result in job loss for those in the military, as well as for those working in related industries

Another potential impact is the psychological effect of autonomous weapons on both soldiers and civilians. The use of autonomous weapons may reduce the emotional toll of combat on soldiers, as they will no longer be directly involved in making decisions that result in the taking of human life. However, the use of these systems could also lead to a desensitization to violence and death, and potentially even contribute to the further devaluation of human life.

Society will also have to grapple with the moral and ethical questions surrounding the use of autonomous weapons. These systems raise questions about accountability and the ability to attribute responsibility in cases of civilian casualties or unintended harm. The use of autonomous weapons could also contribute to an arms race between nations, as each tries to gain an advantage in this new form of warfare.

The impact on international relations must also be considered. The use of autonomous weapons could increase tensions between nations and potentially lead to an erosion of the rules-based international order. It will be crucial for nations to work together to establish clear norms and guidelines for the use of these systems, to ensure that they are used in a responsible and ethical manner.

The long-term effects of autonomous weapons on society are difficult to predict. It will be crucial to carefully consider and address all of the possible ramifications of these developments as we move towards a future where algorithmic warfare becomes increasingly prevalent.

Preparing for the Future of Algorithmic Warfare

THE INTERNATIONAL COMMUNITY must be prepared for the potential impacts of algorithmic warfare in the near future. The development and use of autonomous weapons has the potential to revolutionize how wars are fought and have far-reaching effects on society as a whole. In order to be well-prepared for this future, it is essential to learn about the state of the technology, its possible benefits and risks, and the ongoing debates surrounding the ethical and legal implications of its use.

Education and awareness are among the most important means of preparing for the future of algorithmic warfare. It's not enough to just inform the military and the government; the general public needs to be informed as well. People will be better able to make decisions about the technology's use if they have a thorough understanding of its potential benefits, drawbacks, and risks.

The development of ethical and legal frameworks is another crucial step in getting ready for the future of algorithmic warfare. As technology continues to advance, it will become increasingly important to establish clear guidelines for the ethical and responsible use of autonomous weapons. To accomplish this, the military and civilian experts must work together and actively involve the international community and other stakeholders.

Investment in the research and development of new technologies and capabilities that can help to mitigate the risks posed by autonomous weapons is also essential. The solution could involve anything from investing in the research and development of countermeasures to the threat posed by autonomous weapons to creating new algorithms and systems that are more transparent and secure.

Preparing for the future of algorithmic warfare will require a multi-faceted approach that engages a range of stakeholders, from military personnel and policymakers to experts in technology and ethics. By working together, it is possible to ensure that this new era of conflict is guided by principles of responsibility, transparency, and accountability, and that it ultimately leads to a more secure and stable world.

Unit IX. Conclusion

Summary of Key Findings

THE MAIN FOCUS OF THIS book has been to provide an in-depth examination of the rise of autonomous weapons and the potential benefits and risks associated with their use. Throughout the book, we have explored the history and current state of algorithmic warfare, major advances in technology, international regulations and cooperation, ethical considerations, and the impact of autonomous weapons on society.

Our key findings can be summarized as follows:

1. Algorithmic warfare is rapidly advancing, and autonomous weapons are becoming increasingly capable and sophisticated.
2. The deployment of autonomous weapons has the potential to provide significant benefits in terms of increased efficiency, speed, and reduced collateral damage.
3. However, there are also serious risks associated with the deployment of autonomous weapons, including the possibility of unintended harm, loss of human control, and violations of international law.
4. There is currently a lack of clear national and international regulations for autonomous weapons, and there is a need for greater international

cooperation to mitigate potential risks and ensure responsible and ethical use.

5. The impact of autonomous weapons on society is significant and will likely continue to grow as technology advances.

Overall, it is clear that the rise of algorithmic warfare and the deployment of autonomous weapons is a complex and multi-faceted issue that requires careful consideration and thought. We hope that this book has provided a comprehensive and insightful examination of the subject, and that it will be a valuable resource for those looking to better understand the current state of the field and what the future may hold.

Implications for Policymakers and Military Leaders

AS THE WORLD CONTINUES to grapple with the implications of algorithmic warfare, it is critical for policymakers and military leaders to understand the key findings that have emerged from research, analysis, and practical experience. Let's take a look at the key learnings from our analysis and discuss how these findings might inform future conflict and military planning.

One of the most significant findings of our study is the rapid pace of technological change in the realm of autonomous weapons. In a matter of just a few decades, we have seen the development of increasingly sophisticated algorithms and systems that can carry out complex tasks without human intervention. This has the potential to revolutionize the way that wars are fought, both in terms of how operations are executed, as well as how decisions are made about the use of force.

Another key finding of our study is the need for international cooperation in the development and deployment of autonomous weapons. While it is clear that major global powers such as the US, Russia, and China will play a significant role in shaping the future of this field, it is equally important that they work together to establish norms and standards that can ensure the safe and responsible use of these systems. This will require a deep commitment to dialogue, transparency, and collaboration, even in the face of serious geopolitical challenges.

Policymakers and military leaders must also be prepared to address the many ethical and legal challenges posed by autonomous weapons. This will require an in-depth understanding of the international legal frameworks that govern the use of force, as well as the ethical implications of delegating the power to take life to machines. Leaders must also be mindful of the potential risks and unintended consequences that may arise from the use of these systems, and work to minimize harm to civilians and other non-combatants.

The rise of algorithmic warfare presents policymakers and military leaders with both opportunities and challenges. To be effective, leaders must have an in-depth understanding of the key findings and trends in this field and be ready to address the numerous complex issues that will emerge in the coming years. We can ensure that autonomous weapons play a positive role in shaping the future of warfare and conflict through careful planning and a commitment to ethical and responsible practices.

Final Thoughts on the Future of Autonomous Weapons

THE DEVELOPMENT AND use of autonomous weapons has been a topic of intense discussion and debate in recent years. As the world continues to advance in technology, the possibilities for autonomous weapons and algorithmic warfare seem to be endless. In this final section, we will reflect on the key findings of this book and provide some final thoughts on the future of autonomous weapons.

Autonomous weapons have the potential to revolutionize the way wars are fought, but they also present a number of risks and challenges. The key findings of this book highlight the importance of balancing the risks and benefits of algorithmic warfare and ensuring that the use of autonomous weapons aligns with international law and ethical considerations.

One of the key findings of this book is the need for continued international cooperation in the development and use of autonomous weapons. The global nature of this issue means that it will require the collaboration of multiple nations, international organizations, and experts from a variety of fields to effectively address the challenges posed by algorithmic warfare.

Another important finding is the need for clear and consistent national and international regulations to govern the use of autonomous weapons. These regulations must take into account the unique nature of algorithmic warfare and provide a framework for responsible and ethical use of autonomous weapons.

In terms of the future of algorithmic warfare, it is impossible to predict exactly what form it will take, but it is clear that it will play a significant role in shaping the world of tomorrow. Policymakers and military leaders must be proactive in preparing for the future of algorithmic warfare and be ready to respond to the changing landscape of war and conflict.

The development and use of autonomous weapons must be guided by an ethical commitment and a respect for international law, it is important to note. It is crucial to approach algorithmic warfare with caution and a thorough understanding of its potential risks and benefits, due to its far-reaching effects on society.

The future of algorithmic warfare and autonomous weapons is multifaceted and complex. To ensure that this technology is used responsibly and effectively, continued collaboration, research, and a commitment to ethical considerations will be necessary. The future of algorithmic warfare is uncertain, but we can help shape it in a way that benefits humanity as a whole through careful consideration and preparation.

Unit X. References

Books, articles, and Websites

Books:

- "The Age of AI: And Our Human Future" by by Henry A Kissinger, Eric Schmidt and Daniel Huttenlocher
- "The Future of War: A History" by Lawrence Freedman
- "The Geopolitics of Artificial Intelligence: Strategic Implications of AI for Global Security" by Josh Luberisse
- "The Pentagon's Brain: An Uncensored History of DARPA, America's Top-Secret Military Research Agency" by Annie Jacobsen
- "Lethal Autonomous Weapons: Re-Examining the Law and Ethics of Robotic Warfare (Ethics, National Security, and the Rule of Law)" edited by Jai Galliott, Duncan MacIntosh and Jens David Ohlin
- "Army of None: Autonomous Weapons and the Future of War" by Paul Scharre
- "The Kill Chain: Defending America in the Future of High-Tech Warfare" by Christian Brose
- "Future War" by Christopher Coker
- "T-Minus AI: Humanity's Countdown to Artificial Intelligence and the New Pursuit of Global Power" by Michael Kanaan

- "2034: A Novel of the Next World War" by by Elliot Ackerman and Admiral James Stavridis USN
- "Routledge Handbook of Ethics and War": Just War Theory in the 21st Century" edited by Fritz Allhoff, Nicholas G. Evans, Adam Henschke
- "Drones and Responsibility: Legal, Philosophical and Socio-Technical Perspectives on Remotely Controlled Weapons by Ezio Di Nucci, Filippo Santoni de Sio
- "Armed Drones and the Ethics of War: Military virtue in a post-heroic age" by Christian Enemark

Articles:

- “Lethal Autonomous Weapons Exist; They Must Be Banned It may not be too late to put the evil "Slaughterbots" genie back in the bottle, if the world acts now” by STUART RUSSELLANTHONY AGUIRREEMILIA JAVORSKYMAX TEGMARK, IEEE SPECTRUM
- “An autonomous robot may have already killed people — here’s how the weapons could be more destabilizing than nukes” by James Dawes, C4ISRNE
- “The Risks of Autonomous Weapons Systems for Crisis Stability and Conflict Escalation in Future U.S.-Russia Confrontations” by Burgess Laird, The RAND Blog
- “Defense Innovation Unit chief on keeping the office relevant” by Aaron Mehta, Defense News
- “DIUx Releases Guide For Faster, Less-Costly Technology Acquisition” by Rick Docksai, DOD News
- “Death by algorithm: the age of killer robots is closer than you think” by Kelsey Piper,
- “What is Project Maven? The Pentagon AI project Google employees want out of” by Adam Frisk, Global News

- "Project Maven to Deploy Computer Algorithms to War Zone by Year's End" by Cheryl Pellerin, DOD News
- "Project Maven: Startups Backed By Google, Peter Thiel, Eric Schmidt And James Murdoch Are Building AI And Facial Recognition Surveillance Tools For The Pentagon" by Thomas Brewster, Forbes
- "Intelligence agency takes over Project Maven, the Pentagon's signature AI scheme" by Nathan Strout, C4ISRNET
- "Artificial Intelligence and International Security" by Michael Horowitz, Paul Scharre, Gregory C. Allen, Kara Frederick, Anthony Cho and Edoardo Saravalle
- "UN fails to agree on 'killer robot' ban as nations pour billions into autonomous weapons research" by James Dawes
- "Applying arms-control frameworks to autonomous weapon" by Zachary Kallenborn
- "Stopping Killer Robots: Country Positions on Banning Fully Autonomous Weapons and Retaining Human Control" Brian Stauffer
- "The Third Revolution in Warfare" by Kai-Fu Lee
- " COUNTER-SWARM: A GUIDE TO DEFEATING ROBOTIC SWARMS " by Paul Scharre, War on the Rocks
- " Killer Robots Are Here—and We Need to Regulate Them" by David Cortright and Laura M. Luca, Foreign Policy
- "YES, UNMANNED COMBAT AIRCRAFT ARE THE FUTURE" by Paul Scharre, War on the

Rocks

- "COUNTER-SWARM: A GUIDE TO DEFEATING ROBOTIC SWARMS" by Paul Scharre, War on the Rocks
- "PRECISION-GUIDED WEAPONS COME TO THE INFANTRY" by Paul Scharre, War on the Rocks

Academic Papers:

- “Pros and Cons of Autonomous Weapons Systems” by Amitai Etzioni, PhD and Oren Etzioni, PhD, Military Review, Army University Press
- " Armed Drones and the Ethics of War: Military Virtue in a Post-Heroic Age" by Christian Enemark, Journal of Military and Strategic Studies
- “The Third Offset Strategy and the Army modernization priorities” by Eric P. Hillner, Director’s Action Group
- " An Introduction to Autonomy in Weapon Systems " edited by Paul Scharre and Michael C. Horowitz
- " Autonomous Military Technology: Opportunities and Challenges for Policy and Law" by James Carafano
- "Cyber Warfare and Cyber Terrorism" edited by N.D. Austin and R.A. Popp
- “AI is Shaping the Future of War” by By Amir Husain PRISM Vol. 9, No. 3, National Defense University Press
- “The Future Impact of Artificial Intelligence on Humans and Human Rights” by Steven Livingston and Mathias Risse, Ethics & International Affairs
- “Bias and Discrimination in Artificial Intelligence”

by Jan C. Weyerer and Paul F. Langer, Interdisciplinary Approaches to Digital Transformation and Innovation - Advances in E-Business Research

- "Speculative Human Rights: Artificial Intelligence and the Future of the Human" by James Dawe, Human Rights Quarterly
- "AUTONOMOUS WEAPON SYSTEMS AND INTERNATIONAL HUMANITARIAN LAW: Identifying Limits and the Required Type and Degree of Human–Machine Interaction" BY Vincent Boulanin, Laura Bruun and Netta Goussac
- "Human dignity and lethal autonomous weapon systems: A Christian ethical positioning from a Catholic point of view" BY Elmar Nass, Zeitschrift für Technikfolgenabschätzung in Theorie und Praxis
- "Autonomous Weapons Systems and Meaningful Human Control: Ethical and Legal Issues" by D. Amoroso, G. Tamburrini, Current Robotics Reports
- "Prohibiting Autonomous Weapons: Put Human Dignity First" by Elvira Rosert, Frank G. C. Sauer
- "Governing lethal behavior: Embedding ethics in a hybrid deliberative/reactive robot architecture part I: Motivation and philosophy" by R. Arkin, 3rd ACM/IEEE International Conference on Human-Robot Interaction (HRI)
- "What's So Bad About Killer Robots?" by A. Leveringhaus
- "An Ethical Governor for Constraining Lethal Action in an Autonomous System" by R. Arkin,

Patrick Ulam and Brittany A. Duncan

- "Robots and Respect: Assessing the Case Against Autonomous Weapon Systems" by R. Sparrow
- "The Dehumanization of International Humanitarian Law: Legal, Ethical, and Political Implications of Autonomous Weapon Systems" by Markus Wagner
- "The Ethics & Morality of Robotic Warfare: Assessing the Debate over Autonomous Weapons" by Michael C. Horowitz
- "Risks and Benefits of Autonomous Weapon Systems: Perceptions among Future Australian Defence Force Officers By Drs. Jai Galliott & Austin Wyatt, Journal of Indo-Pacific Affairs, Air University Press
- "The Novel of Human Rights" by James Dawes, Human Rights Quarterly, Johns Hopkins University Press
- "Autonomous Weapon Systems and the Law of Armed Conflict" by Tim McFarland, University of New South Wales, Canberra
- "Drones, Ethics and the Armchair Soldier" by John Kaag, The Journal of Military Ethics
- "The Use of Unmanned Aerial Vehicles in Contemporary Conflict: A Legal and Ethical Analysis" by Sarah Krepps and John Kaag, University of Chicago Press
- " The Art of Law in the International Community. By Mary Ellen O'Connell. Cambridge, UK: Cambridge University Press

- "The International Law of Drones" by Mary Ellen O'Connell, The American Journal of International Law edited by Noel Sharkey and Michael Nagenborg
- "The Moral Case for the Development and Use of Autonomous Weapon Systems" by Erich Riesen, Journal of Military Ethics
- "Ungoverned Spaces: Alternatives to State Authority in an Era of Softened Sovereignty" by Anne L. Clunan and Harold A. Trinkunas, Journal of Military and Strategic Studies
- "Remote Killing and the Ethics of Drone Warfare" by James DeShaw Rae
- "Leadership Challenges from the Deployment of Lethal Autonomous Weapon Systems: How Erosion of Human Supervision Over Lethal Engagement Will Impact How Commanders Exercise Leadership" by Paddy Walker
- "AI & Global Governance: AI and Nuclear Weapons – Promise and Perils of AI for Nuclear Stability" by Dr Vincent Boulanin, Centre for Policy Research, United Nations University.

Websites:

- "Algorithmic warfare is coming. Humans must retain control" (https://www.weforum.org/agenda/2018/09/algorithmic-war-is-coming-humans-must-retain-control)
- "Project Maven" (https://www.darpa.mil/program/project-maven)
- "Third Offset Strategy" (https://www.defense.gov/Our-Story/Special-Reports/Third-Offset-Strategy/)
- "International Committee of the Red Cross: Autonomous Weapons" (https://www.icrc.org/en/topic/autonomous-weapons)
- "Human Rights Watch: Autonomous Weapons" (https://www.hrw.org/topic/arms/autonomous-weapons)

Reports and Studies

The development and use of autonomous weapons have been widely studied and discussed by various organizations, think tanks, and academic institutions. Here are some of the most significant reports and studies related to algorithmic warfare:

1. DOD DIRECTIVE 3000.09 AUTONOMY IN WEAPON SYSTEMS: This report was created by the United States Department of Defense (DoD). The directive establishes DOD policy and assigns responsibilities for the development and use of autonomous and semiautonomous functions in weapon systems, such as remotely or onboard-operated armed platforms. It also establishes DOD guidelines designed to reduce the likelihood and consequences of failures in autonomous and semi-autonomous weapon systems that could result in unintended engagements.
2. Project Maven Report: This report was created by the Defense Innovation Unit Experimental (DIUx) in the United States Department of Defense (DoD). The report outlines the development of Project Maven, a military project aimed at developing machine learning algorithms to help military analysts process large amounts of aerial drone footage.

1. The Future of Warfare: Autonomous Systems: This report was published by the Royal United Services Institute (RUSI), a British think tank focused on defense and security issues. The report analyzes the future of autonomous systems in warfare and their potential impact on military operations and strategy.

1. Autonomous Weapons and Operational Risk: This study was conducted by the RAND Corporation, a nonprofit global policy think tank. The study

analyzes the risks associated with autonomous weapons, including the potential for unintended consequences and the need for human oversight.

1. The Ethics of Autonomous Weapons: This report was published by the Carnegie Council for Ethics in International Affairs, a leading international ethics organization. The report discusses the ethical considerations surrounding autonomous weapons, including the role of human decision-makers and the potential for unintended harm.

1. Autonomous Weapons and International Law: This report was published by the International Committee of the Red Cross (ICRC), the leading humanitarian organization focused on the protection of civilians in armed conflict. The report discusses the legality of autonomous weapons under international law and the potential for human rights violations.

These reports and studies provide valuable insights into the current state of algorithmic warfare and the potential impact of autonomous weapons on society and international relations. Policymakers and military leaders can use these resources to inform their decisions on the development and deployment of autonomous weapons.

Did you love *Algorithmic Warfare: The Rise of Autonomous Weapons*? Then you should read *The Art of War in the 21st Century: Timeless Principles for Modern Military Strategy*[1] by Josh Luberisse!

[2]

In "The Art of War in the 21st Century," visionary author Josh brings a fresh perspective to the ancient wisdom of Sun Tzu's renowned treatise. Drawing upon his experience as an entrepreneur and his passion for strategic thinking, Josh explores the timeless principles of "The Art of War" and their applications to modern military strategy.

1. https://books2read.com/u/bPLQwj

2. https://books2read.com/u/bPLQwj

This book is not a mere translation or interpretation of Sun Tzu's work. Instead, it serves as a bridge between the ancient wisdom of the past and the complex challenges of the present. Josh's unique perspective combines military strategy with insights from the world of business and innovation, offering a fresh take on how these timeless principles can be adapted and applied in the contemporary landscape.

Through engaging narratives, real-world examples, and thought-provoking analysis, Josh demonstrates how Sun Tzu's principles can guide leaders in the 21st century to navigate the complexities of modern warfare. He explores topics such as understanding the operational environment, harnessing technological advancements, fostering effective leadership, and building resilient coalitions.

"The Art of War in the 21st Century" is not solely aimed at military professionals. It is a book for visionary thinkers, entrepreneurs, and leaders from various fields who seek to enhance their strategic acumen and decision-making abilities. It serves as a guidebook for those navigating the ever-evolving landscapes of business, politics, and conflict resolution.

Josh's fresh approach to Sun Tzu's timeless wisdom makes this book an invaluable resource for those seeking innovative strategies to overcome challenges, seize opportunities, and achieve success in the modern world. The author's ability to bridge the gap between ancient principles and contemporary contexts provides readers with practical and thought-provoking insights that transcend traditional boundaries.

Whether you are a military strategist, a business leader, or an aspiring entrepreneur, "The Art of War in the 21st Century" offers a compelling exploration of strategic thinking in our rapidly changing world. Join Josh on this transformative journey and unlock the secrets of Sun Tzu's wisdom, paving the way for strategic brilliance and remarkable achievements in the 21st century.

About the Author

Josh is a data scientist and a prolific author of books on artificial intelligence, geopolitics, finance and cybersecurity. With several books on these subjects under his belt, Josh has established himself as a thought leader in the field. His most recent books, "Silent Wars: Espionage, Sabotage, and the Covert Battles in Cyberspace", "Silicon and Rare Earth: The Global Contest for Semiconductor and Rare Earth Supremacy", AGI Architects: Building a Symbiotic Civilization with Superintelligent Systems", "Algorithmic Warfare: The Rise of Autonomous Weapons", and "The Geopolitics of Artificial Intelligence: Strategic Implications of AI for Global Security" highlight his deep understanding of the historically unprecedented impact that artificial intelligence is poised to have on the future of the human race. His books have been praised for their practical advice and actionable insights, making them must-reads for anyone looking to gain a better understanding of the implications of emerging technologies for our society, our environment, and our future.

In addition to being an author, Josh is the host of Innovate Now: The Pulse of Future Technologies, a podcast that explores the fascinating world of emerging technologies and the future of innovation by delving into the most groundbreaking advancements in artificial intelligence, fintech, distributed and quantum computing, the AI arms race, and the tech start-up ecosystem. Josh is also a seasoned practitioner. He is the founder of a number of startup companies including one that is focused on AI research and another that provides cybersecurity consulting services, and performs penetration testing engagements.

Josh is committed to staying at the forefront of the rapidly evolving field of emerging technologies. His expertise and passion have made him a sought-after speaker and consultant, and his contributions have been widely recognized and appreciated. Whether through his books, his work with clients, or his field assignments, Josh continues to inspire and educate others in the hope of making a lasting impact on the world of artificial intelligence and cybersecurity. If you're looking for practical and actionable insights be sure to check out Josh's other manuscripts and join his community of readers and followers.

www.ingramcontent.com/pod-product-compliance
Lightning Source LLC
LaVergne TN
LVHW090420210926
848480LV00002B/636
9798223471912